Nightmare Bugs!

Goliath Bird-Eating Spiders

By Natalie Humphrey

Gareth Stevens Publishing

Hot Topics

Please visit our website, www.garethstevens.com. For a free color catalog of all our high-quality books, call toll free 1-800-542-2595 or fax 1-877-542-2596.

Library of Congress Cataloging-in-Publication Data
Names: Humphrey, Natalie, author.
Title: Goliath bird-eating spiders / Natalie Humphrey.
Description: Buffalo, NY : Gareth Stevens Publishing, [2025] | Series: Nightmare bugs! | Includes bibliographical references and index.
Identifiers: LCCN 2023044126 | ISBN 9781538292211 (library binding) | ISBN 9781538292204 (paperback) | ISBN 9781538292228 (ebook)
Subjects: LCSH: Theraphosa blondi–Juvenile literature. | Tarantulas–Juvenile literature.
Classification: LCC QL458.42.T5 H86 2025 | DDC 595.4/4–dc23/eng/20231026
LC record available at https://lccn.loc.gov/2023044126

First Edition

Published in 2025 by
Gareth Stevens Publishing
2544 Clinton St
Buffalo, NY 14224

Designer: Andrea Davison-Bartolotta
Editor: Natalie Humphrey

Photo credits: Cover, p. 23 reptiles4all/Shutterstock.com; series art (texture) Video_ Stock _Production/Shutterstock.com; series art (lines) Hananta S/Shutterstock.com; series art (sign) Mr Doomits/Shutterstock.com; series art (trees) Wilqkuku/ Shutterstock.com; p. 5 Allmy/Shutterstock.com; p. 7 PetlinDmitry/Shutterstock.com; p. 9 juerginho/Shutterstock.com; p. 10 Mehgan Murphy/rawpixel.com/Public Domain Animal Photography/Flickr; p. 11 Panther Media GmbH/Alamy Stock Photo; pp. 13, 19, 21 Milan Zygmunt/Shutterstock.com; p. 14 Saad315/Shutterstock.com; p. 15 The Reptilarium/Flickr; p. 17 Brian Gratwicke/Flickr; p. 25 Juniors Bildarchiv GmbH/ Alamy Stock Photo; p. 27 Bernard DUPONT/Encyclopedia of Life Images/Flickr; p. 29 Wirestock Creators/Shutterstock.com; p. 30 Tobias Hauke/Shutterstock.com.

Printed in the United States of America

CONTENTS

Huge SPIDERS

The Goliath bird-eating spider is the heaviest **arachnid** in the world. This huge spider can weigh up to 6 ounces (170 g). That's as heavy as a hockey puck. This monster spider looks like it's straight out of a nightmare!

Terrible Truths

Goliath bird-eating spiders are tarantulas. Tarantulas are large, hairy spiders that are found in warm places around the world.

Spider BODIES

Goliath bird-eating spiders have eight long legs. From the end of one leg to the end of another leg, this spider can be 11 inches (28 cm)! Their bodies are usually brown in color. The Goliath bird-eater has a large **abdomen**.

Terrible Truths

Spiders have strange bodies. Their head and **thorax** are one part! This freaky body is called a cephalothorax.

Home Sweet HOME

Goliath bird-eaters are found in the rainforests of northern South America. They are usually found in Venezuela, northern Brazil, Guyana, French Guiana, and Suriname. They like damp, or somewhat wet, and swampy ground with plenty of spots for them to hide.

Terrible Truths

Goliath bird-eaters can only be found in rainforests, which are areas with many trees and a lot of rain.

Goliath bird-eating spiders live in burrows. Like most spiders, Goliath bird-eaters make silk in their body. This silk is strong, thin, and looks like thread. Goliath bird-eaters use their silk to cover the inside of their home.

Terrible Truths

A burrow is a hole an animal digs and uses as their home. Goliath bird-eaters dig their own holes or will take over a home left behind by another animal.

Dinner TIME!

Goliath bird-eating spiders will eat any animal smaller than them. They usually eat mice, snakes, lizards, and bugs. Their name says they eat birds, but they don't do this often. While Goliath bird-eaters are big enough to eat birds, they can't usually catch them!

Terrible Truths

A Goliath bird-eating spider has huge **fangs** that are about 1 inch (2.5 cm) long!

SURPRISE!

Goliath bird-eating spiders are ambush hunters. This means a Goliath bird-eater will wait for its dinner to come close. Then, the spider will sink its fangs into its prey. Prey is an animal that is eaten by another animal as food.

Terrible Truths

A Goliath bird-eating spider's fangs are strong enough to break through a mouse's **skull**.

Venomous BITES

A Goliath bird-eating spider's bite is **venomous**. When a Goliath bird-eater bites its prey, its venom will paralyze the animal. This means the animal is unable to move! After its prey stops moving, the Goliath bird-eating spider pulls its dinner into its burrow.

Terrible Truths

Not every spider bite is venomous. Goliath bird-eating spiders can sometimes dry bite. A dry bite is when a venomous animal bites without **injecting** venom.

Spiders don't chew their food. When the Goliath bird-eater is safe in its burrow, it injects **enzymes** into its prey. These enzymes break down the insides of the spider's prey! After the prey's insides are liquefied, the Goliath bird-eating spider drinks them!

Terrible Truths

Liquefying is the process of turning something into a liquid, somewhat like water.

Stay BACK!

The Goliath bird-eating spider has a freaky way to keep itself safe. If a Goliath bird-eater can't get away from danger, it makes a hissing sound by rubbing its first and second pairs of legs together to scare predators away. If that doesn't work, the spider attacks!

Terrible Truths

Goliath bird-eating spiders have eight eyes, but they can't see very well! They use the hair on their legs and abdomen to help them sense when danger is close by.

Hairy ATTACK

When a predator won't leave a Goliath bird-eating spider alone, this spider has a special attack! Rubbing its legs together, the spider can fire its hair at a predator. These little hairs get stuck in the predator's skin and can hurt or itch!

Terrible Truths

A Goliath bird-eating spider's hair has barbs, or sharp, backward-facing points. This makes it so that when the hair sticks into a predator, it's hard to get out.

So Many EGGS

Female Goliath bird-eating spiders can lay a lot of eggs! One female spider can lay up to 150 eggs at the same time. She wraps her eggs in her own silk. She makes a ball-like egg sack. Then, the female keeps the egg sack in her burrow and guards it.

Terrible Truths

The egg ball the female spins can be almost as big as a tennis ball!

When the baby spiders come out of their eggs, they stay close to their mother for two to three years. As a Goliath bird-eating spider grows, it sheds its skin! This is called molting. A Goliath bird-eating spider molts several times before it is fully grown.

Terrible Truths

After a spider molts, its skin is soft for a short period of time. Then, the spider's skin hardens and creates a shell called an exoskeleton.

shed skin
spider

Painful BITES

Goliath bird-eaters usually stay far away from humans. But if a human picks up a Goliath bird-eater, the spider might bite that person! Goliath bird-eating spider bites aren't deadly to humans, but they do hurt. These spiders should be given a lot of space.

Terrible Truths

Goliath bird-eaters might be dangerous, but some people in South America think they make a tasty dinner. These people cook them with banana leaves and say they taste like shrimp!

Tiny Terrors

GOLIATH BIRD-EATING SPIDER

Size:	leg-span of 11 inches (28 cm)
Appearance:	dark brown or reddish and hairy
Life span:	males usually live three to six years; females often live 15 to 25 years
Where Do They Live?	the rainforests of northern South America
Diet:	bugs, mice, and anything small enough for it to catch
Nightmare Fact:	A Goliath bird-eater's hiss can be heard up to 15 feet (4.6 m) away.

FOR MORE INFORMATION

BOOKS

Archer, Claire. *Bird-Eating Spiders.* Mankato, MN: Abdo Kids Jumbo, 2021.

Llewellyn, Claire. *Scary Spiders.* New York, NY: Kingfisher, 2021.

WEBSITES

San Diego Zoo Wildlife Explorers: Goliath Bird-Eating Spider
www.sdzwildlifeexplorers.org/animals/Goliath-bird-eating-spider
Check out more photos and learn more about how Goliath bird-eating spiders live in the wild.

Smithsonian's National Zoo: Goliath bird-eating tarantula
nationalzoo.si.edu/animals/goliath-bird-eating-tarantula
Take an up-close look at the Goliath bird-eating spider and learn more about them.

GLOSSARY

abdomen: The final part of an insect's body that includes most of its guts and organs for mating.

arachnid: One of a large class of small animals that includes spiders, scorpions, ticks, daddy longlegs, and mites.

enzyme: Matter made in the body that helps certain actions necessary for life to occur.

fang: A long, pointed tooth.

inject: To use sharp teeth to force venom into an animal's body.

skull: The bony frame of the head and face.

thorax: The middle part of an insect's body. The wings and legs connect to it.

venomous: Able to produce a liquid called venom that is harmful to other animals.

INDEX